Intel® Media Processor SDK

A Quick Start Guide

David Kauer

GlassThumb Inc.

Intel Media Processor SDK: A Quick Start Guide

By David Kauer

ISBN: 978-1-105-43454-9

Copyright © 2011 GlassThumb Inc. All Rights Reserved.

Published by GlassThumb Inc. 817 Broadway, New York, NY 10005

Printing History:

November 2011:   First Edition.

Many of the designations used by manufactures and sellers to distinguish their products are claimed as trademarks. Where those designations appear in this book, and GlassThumb Inc. was aware of a trademark claim, the designations have been printed in caps or initial caps.

While ever precaution has been taken in the preparation of this book, the publisher and author assume no responsibility for errors or omissions, or for damages resulting from the use of the information contained herein.

0-212-006218-1

Table of Contents

I. Introduction 1

The Intel® Media Processor CE4100 Pocket Guide 1

What is the CE4100 Media Processor? 1

Development Platform Producers and High-Level Features 2

Starting Tools You Will Need 3

Quick Overview of Linux and Embedded Linux 4

Why Choose the CE4100 Media Processor? 5

II. Overview of Intel® Media Processor CE4100 7

Brief Description of Development Platform Contents 7

Development Platform Hardware Overview 9

Components on the Development Platform 12

Development Utilities and Supporting Libraries for Intel® CE Media Processor SDK: 14

Detailed Development Platform Contents 18

BusyBox Commands 18

BUSYBOX COMMAND LIMITATIONS 31

III. Hooking up to the Development Platform 35

SETUP TOOLS FOR THE DEVELOPMENT PLATFORM 35

SERIAL PORT TO USB/SERIAL PORT INFORMATION 36

IV. Connecting the Videon Central AV platform to a TV display 38

TRANSFERRING FILES FROM HOST TO TARGET AND VICE VERSA 39

HELPFUL PLATFORM COMMANDS 41

V. Setting Up and Understanding the SDK 42

THE SDK DEVELOPMENT ENVIRONMENT 46

THE TARGET FILE SYSTEM 48

DEBUGGING 49

VI. Samples, Applications, and Scripts for CE4100 Platform 51

INFORMATION ON GDL, GRAPHICS, DIRECTFB SAMPLES 51

INFORMATION ON MEDIA APPLICATIONS 60

SAMPLE CONFIGURATION FILES FOR MEDIA APPLICATIONS 69

DEVICE MEMORY MODIFICATION FOR DISPLAY/SMD 76

VII. Script Information and Samples 78

VIII.   SDK Development Setup and Custom Applications 81

SDK Development Setup 81

Basic Applications Running on CE4100 Console 83

Advanced Applications Running on TV Display 84

IX. RedBoot and CEFDK Shell 88

Redboot Introduction 88

CEFDK Shell and Commands 88

Redboot Commands 90

Modifying the Redboot Script 92

X.   Appendix A: Host Environment Setup 97

XI. Appendix B: Online Resources 99

XII. Glossary 100

I. Introduction

The Intel® Media Processor CE4100 Pocket Guide

The purpose of this pocket guide is to bridge the gap between desktop and embedded Linux development. It provides additional background information to the Intel CE4100 SDK documents; especially to those who are new or have little experience in embedded Linux environments. This pocket guide is intended for readers who are interested in developing for the Intel Media Processor CE4100 Development Platform. There are a few items to note about this guide: for starters, this guide is based upon a Development Platform with PR 13 installed and the Platform does not have an integrated Blu-ray Disc player. Therefore, this guide is not meant for developers who are solely interested in Blu-ray devices and/or technology. In addition, this guide will slant towards using a Linux Host, and it is recommended that the user has a Linux Host PC in order to connect to the Development Platform. Please note that some steps and commands in this guide may require you to have root privileges.

What is the CE4100 Media Processor?

The Intel® Media Processor CE4100 Development Platform is a high performance tool and is meant for developers who are interested in the embedded environment. It is specifically designed to provide convenience for developers, and promotes development for software applications. There are many open-source components on the Platform for developers to control and manipulate such as audio/video applications and samples. The Development Platform is operated by a distribution of Linux, which can run set-top box applications and other developments. Intel's® Atom Processor CE4100 SOC (System-on-Chip) is used to power the Development Platform.

Development Platform Producers and High-Level Features

Intel® is responsible for producing the Intel® Media Processor CE4100 Development Platform as well as the Intel® Atom Processor. These devices are able to bring the Internet from personal computers to TV displays using Intel® Architecture (IA), which is a great asset for CE device developers. One of the development platforms for the CE 4100 is sold by Videon Central. Videon Central provides digital video solutions for consumer electronics. They offer a number of services such as software development, system design and integration, and product testing for equipment manufacturers and System-on-Chip (SoC) organizations such as Intel®.

The CE4100 Development Platform offers a number of high-level features:

- Enhances the capabilities and performance of media applications
- Five universal pixel planes (UPP)
- Two HD video decoders
- HDMI video output/input
- Powerful 3D graphics and video renderer
- Minimal power usage
- Audio/Video samples are provided
- Security Processor

Starting Tools You Will Need

There are a few basic items that are needed in order to start developing with the Intel® CE Media Processor CE4100. Most, if not all, of the items below may have come with the Platform or can be easily purchased online.

- Female RS-232 cable or a male RS-232 to USB adapter for remote access
- HDMI or Component and Audio cable for video and audio output
- DC power supply
- Ethernet cable for network access

Quick Overview of Linux and Embedded Linux

Linux is a free and popular operating system of choice and is known throughout the world. Linux can be found on many embedded systems including DVD players, smart devices, digital cameras, network switches, etc. You may already own a Embedded Linux device without even knowing it. Embedded Linux is used in millions of cell phones and smart devices that are distributed worldwide. Many companies and corporations customize their own version of embedded Linux to allow for easier accessibility and utilization.

There are a few reasons why embedded Linux has become very successful:

- Linux has support for numerous hardware devices
- Linux has application and networking protocol support
- Linux can range from small consumer electronics to large switch networks and routers
- A growing number of hardware and software vendors support Linux
- Many active developers are interested in Linux, which allows for quick and reliable device, hardware and platform support.

We are seeing a high adoption rate in household items for Linux, scaling from your HDTV to your smart phone device. Examples of Linux embedded systems can be found in DVD players, Android Tablets, video games platforms, digital cameras, wireless networking hardware, network switches and more. No one can neglect or disregard the increasing presence of embedded Linux in the marketplace today.

Why Choose the CE4100 Media Processor?

Using the Intel® architecture in CE devices provides developers with several benefits. One of these benefits includes a development community of over 5000 software vendors. Developers will also gain from solid tool-chains and development support through Intel® and other vendors.

Videon Central is dedicated the platform and has an expert development team to assist companies and developers. They work with a variety of new and growing technologies such as Blu-ray CE manufactures 3D entertainment and in-flight media systems. With nearly 15 years since its founding Videon Central has a number of expert employees that provide excellent customer service.

II. Overview of Intel® Media Processor CE4100

Brief Description of Development Platform Contents

The file system inside the development platform contains a number of directories. You will probably become more familiar with bin, etc, and usr as these directories contain media applications, samples, or configuration files. The following provides a brief description for each directory:

bin - A collection of programs that may be used by the system, administrator and users. The bin includes the GDL and GFX samples as well as a several applications such as cl_app and sg_player.

bzImage - Contains the startup files and the Linux Kernel Image. See Redboot scripts.

dev - Contains references to the peripheral hardware on the CPU, which are represented as device files in the form of characters, blocks, sockets, and links.

etc - Configuration files that are for the system and are specified by the host. It includes directFB and platform configuration files.

hdisk1 - Contains the contents of the disk inserted into the Development Platform. This directory is useless if the Platform did not come with an integrated Blu-ray Disc Player.

lib - Library files that are used for programs needed by the system and users.

linuxrc - This is a program that starts prior to boot up. It allows you manually load drivers that are required as modules. It links to bin/busybox.

mnt - Mount point for external file systems such as a CD/DVD or digital camera.

proc - A virtual file system that contains information about the Linux kernel. It describes the currently-running processes in the kernel. Many of the files in this directory are used by programs.

sbin - Programs that are meant to be used by the system and administrator.

scripts - Ideal location to put user-made script files.

share - Program-specific files, such as examples and installation instructions.

sys - This directory includes the Firmware, Kernel, and system files.

tmp - Temporary space used by the system. Any data stored here is cleared upon reboot.

usr - Contains user-related programs such as applications, libraries, documentation, etc. This includes directFB samples and diagnostic applications such as graphics and memory.

var - Storage for user-created temporary and variable files. These files are specific to the computer, and it is created and updated as it runs. It includes logs, mail queues, temporary Internet files, a CD image, etc.

Development Platform Hardware Overview

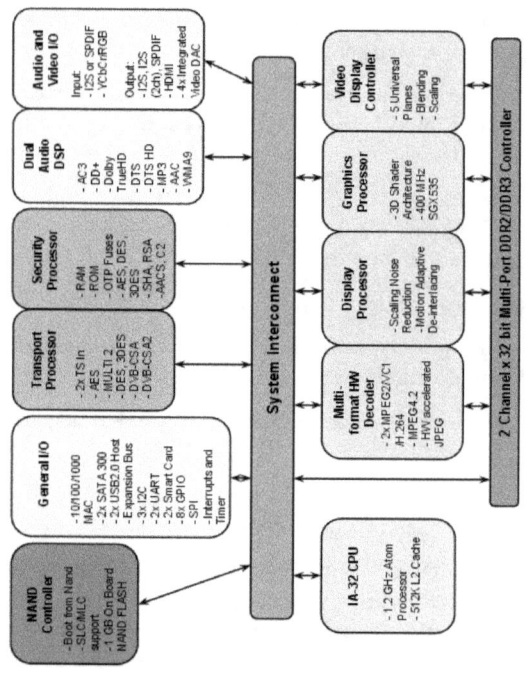

Figure 1 - Architecture of Intel® Atom Process

Technical Specifications:

System & Hardware	Operating System: Linux
	Intel® Atom Processor CE4100 • 1.2 GHz Atom Processor • 512K L2 Cache
	Hardware-Based Security Processor
	400 MHz SGX535 Graphics Processor
	Display Processor • Scaling • Noise Reduction • Motion Adaptive De-Interlacing • Supports Output Resolutions Up To 1080P60
	1 GB DDR-3 SDRAM
	1 GB On Board NAND FLASH
Video Output	HDMI/Component Video Output

	Multi-Format HW Video Decoder (Video Output) • 2 x MPEG-2/VC1/H.264 • 1 x MPEG4.2/DivX • HW Accelerated JPEG
Audio Output	Dual Audio DSP • 533 MHz • MPEG and AAC
	Audio Outputs • L/R Analog Audio • HDMI • SPDIF (Optical and Coaxial)
Inputs	2 HDMI Inputs
	USB Connector for USB 2.0 or Type A
	RS-232 Connector: DB-9 Male
	RJ45 (10/100/1000) Ethernet Connector

Environmental Specifications:

Qualifications	Operating temperature of 0 to 50 degrees C
	Storage temperature of -20 to 70 degrees
	5% to 90% relative humidity
	12 VDC Power Supply

Components on the Development Platform

GCC compiler version 4.1.2: GNU Compiler Collection allows the user to compile Linux applications for the Development Platform on the host system.

glibc version 2.7 (packaged with gcc 4.1.2): glibc defines system calls and other system facilities using C libraries.

binutils 2.17.50.0.17: binutils handles binary object files through GNU linker, GNU Assembler, and other utilities.

GDB - Linux x86 version 6.8: GDB is the GNU Project debugger. This will allow you to see what happened to a program when it reaches an error.

RedBoot* boot loader 2: RedBoot or RedHat Embedded Debug and Bootstrap Program is designed for the embedded operating system bootstrap/loader.

Linux* kernel 2.6.23 or Linux* kernel 2.6.28: The Linux kernel is the heart of a Linux system and is responsible for accessing the computer's processes and hardware components.

BusyBox Environment 1.10.2: BusyBox provides a variety of many common UNIX functions and is a small shell that is run in the Linux kernel. Refer to BusyBox commands and limitations.

Nandflash: The Nandflash can be used as a storage device and is known to be a Linux block driver.

ALSA (Advanced Linux Sound Architecture) x86: ALSA or Advanced Linux Sound Architecture provides support for many sound devices through advanced utilities.

DirectFB* 1.2: DirectFB supports input devices, windowing, and grapics, and is an Open Source API.

Development Utilities and Supporting Libraries for Intel® CE Media Processor SDK:

- libjpeg
- libpng
- libz
- FreeType 2.3.9
- ncurses 5.6
- libiconv 1.11.1
- libusb 0.1.12
- libmtp 0.2.6
- udev 120
- libnl
- dhcpv6

The source code for each of the open source components can be found in the `/IntelCE-<version>/project_build_i686/IntelCE/devtools-<version>/` directory.

NTFS-3G-Linux-x86: Allows the use of NTFS file system through ntfs-3g support. The following commands will mount and unmount the NTFS in read/write mode:

```
ntfs-3g /dev/sda1 /mnt
```

```
umount /mnt
```

Other Open Source Components:

- adduser
- autoconf 2.59
- automake 1.9.6
- Bison 2.3
- chgrp
- chown
- chvt
- crond
- delgroup
- deluser
- dumpkmap
- fbset
- fdflush
- fdformat
- Flex 2.5.35
- fsck
- fsck.minix
- GStreamer – Linux x86 version 0.10.17
- httpd
- Htuple – Linux x86
- ifdown

- ifup
- 18
- ipclib-Linux-x86
- iptunnel
- kdb_mode
- libcurl-Linux-x86
- loadfont
- loadkmap
- login
- logname
- m4 1.4.9
- mdev
- mkfs.minix
- modprobe
- mt
- NSPR-Linux-x86
- ntfs-3g
- OpenGL* ES 1.1 / 2.0
- OpenSSL*-Linux-x86
- OpenVG* 1.0
- passwd
- patch 2.5.4
- pivot_root

- pscan
- raidautorun
- readprofile
- resize
- rpm
- runlevel
- setkeycodes
- setuidgid
- SGX graphics kernel module
- su
- sulogin
- switch_root
- sysctl
- system_utils-Linux-x86
- toolchain-Linux-x86
- UDF patch 2.5
- udhcpd
- vconfig
- vlock
- watchdog
- who
- zlib-1.2.3
- libtool-1.5.26

Detailed Development Platform Contents

This section will display some of the more important files and directories inside the Development. The contents will range from sample applications to media examples to configuration files.

BusyBox Commands

The following commands are included on the Intel® CE Media Processor CE 4100 Development Platform though busybox:

- addgroup - Add a group
- adjtimex - Read and set system timebase parameters
- ar - Extract or list files from an ar archive
- arp - Manipulate the ARP cache
- arping - Send ARP requests or replies
- ash - A small shell, usually the default
- awk - Data extraction and reporting tool
- basename - Strip directory path and suffixes from a file
- bunzip2 - Uncompress a file

- bzcat - Uncompress to stdout
- bzip2 - Compress a file with bzip2 application
- cal - Display a calendar
- cat - Concatenate a file and print them to stdout
- catv - Display nonprinting characters as ^x or M\-x
- chattr - Change file attributes on an ext2 file system
- chmod - Change permissions on a file
- chpasswd - Read user: password information from stdin and update /etc/passwd accordingly
- chpst - Change the process state and run program
- chroot - Run command with root directory set to Newroot
- chrt - Manipulate real-time attributes of a process
- cksum - Calculate the CRC32 checksums of a file
- clear - Clear terminal screen
- cmp - Compare file1 to stdin if file2 is not specified
- comm - Compare file1 to file2, or to stdin if specified
- cp - Copy source file to destination file

- cpio - Extract or list files from a cpio archive, or create a cpio archive
- cut - Print selected fields from each input file to standard output
- date - Display time or set time
- dc - Tiny RPN calculator
- dd - Copy a file with converting and formatting
- deallocvt – De-allocate unused virtual terminal /dev/ttyn
- df - Print file system usage statistics
- diff - Compare files line by line and output the differences
- dirname - Strip a non directory suffix from a filename
- dmesg - Print or control the kernel ring buffer
- dos2unix - Convert a file from DOS to Unix format
- du - Summarize disk space used for each file and/or directory
- echo - Print the specified ARGs to stdout
- ed - Eject a specified device (default /dev/cdrom)
- egrep - Extended regular expression syntax
- eject - Eject a specified device (or default /dev/cdrom)

- env - Print the current environment or run a program after setting
- envdir - Set various environment variables as specified by files
- envuidgid - Set User ID to account's User ID and Group ID to account's Group ID and run program.
- expand - Convert tabs to spaces, writing to standard output
- expr - Print the value of an expression to standard output
- false - Return an exit code of 'false'
- fdisk - Change partition table
- fgrep - Search for files
- find - Search for files. The default PATH is the current directory, default expression is '-print'
- fold - Wrap input lines in each file (standard input by default)
- free - Display the amount of free and used system memory
- freeramdisk - Free all memory used by the specified ramdisk
- ftpget - Retrieve a remote file via FTP
- ftpput - Store a local file on a remote machine via FTP
- fuser - Find processes that use files or ports

- getopt - Parsing command-line arguments
- getty - Open a tty, prompt for a login name, and then invoke /bin/login
- grep - Search for a pattern in each file or standard input
- gunzip - Uncompress a file (or standard input)
- gzip - Compress a file (or standard input)
- halt - Halt the system
- hdparm - Get/set HD device parameters
- head - Print first ten lines of each file to standard output
- hexdump - Display files or standard input in a user-specified format
- hostid - Print a unique 32-bit identifier for the machine
- hostname - Get or set hostname or DNS domain name
- hwclock - Query and set hardware clock (RTC)
- id - Print information about user or the current user
- ifconfig - Configure a network interface
- inetd - Listen for network connections and launch programs
- init - init is the parent of all processes

- insmod - Load the specified kernel modules into the kernel
- install - Copy files and set attributes
- ip - Show/manipulate routing, devices, policy routing, and tunnels
- ipaddr - ipaddr {add | delete} IFADDR dev String
- ipcalc - Calculate IP network settings from an IP address
- ipcrm - Uppercase options MQS remove an object by shmkey value
- ipcs - Provide information on ipc facilities
- iplink - iplink set Device {up | down | arp | multicast | on | off}
- iproute - iproute {list | flush| Selector
- iprule - iprule {list | add | del} Selector action
- kill - Send a signal (default is TERM) to given PIDs
- killall - Send a signal (default is TERM) to given processes
- killall5 - Send a signal (default is TERM) to all processes outside the current session
- klogd - Kernel logger
- length - Print String's length
- less - View a file or list of files

- linux32 - Create a link to the specified target
- linux64 - Create a link to the specified target
- linuxrc - Create a link to the specified target
- ln - Create a link to the specified target
- logger - Write message to the system log
- logread - Show messages in syslogd's circular buffer
- losetup - Set up and control loop devices
- ls - List directory contents
- lsattr - List file attributes on an ext2 fs
- lsmod - List the currently loaded kernel modules
- lzmacat - Uncompress to stdout
- makedevs - Create a range of special files as specified in a device table
- md5sum - Print or check MD5 checksums
- mesg - Control write access to your terminal
- microcom - Copy bytes for stdin to TTY and from TTY to stdout
- mkdir - Create a directory
- mkfifo - Create a named pipe (identical to mknod name p)
- mknod - Create a special file (block, character, or pipe)

- mkswap - Prepare a block device to be used as a swap partition
- mktemp - Create a temporary file
- more - View a file or standard input one screen at a time
- mount - Mount a file system
- mountpoint - Check if the directory is a mountpoint
- mv - Rename source to destination
- nameif - Rename the network interface while it's in the down state
- nc - TCP/IP Swiss army knife
- netstat - Display networking information
- nice - Run a program with a modified scheduling priority
- nmeter - Monitor the system in real time
- nohup - Run a command immune to hangups, with output to a non-tty
- nslookup - Query the name server for the IP address of the given host
- od - Write an unambiguous representation of a file
- openvt - Start command on a new virtual terminal
- patch - Apply a diff file to an original

- pgrep - Display processes selected by the regex pattern
- pidof - List PIDs of all processes with names that match NAMES
- ping - Send packets to network hosts
- ping6 - Send packets to network hosts
- pipe_progress - Move the current root file system
- pkill - Send a signal to processes selected by the regex pattern
- poweroff - Halt and shut off power
- printenv - Print all or part of the environment
- printf - Format and print arguments
- ps - Report process status
- pwd - Print the full filename of the current working directory
- rdate - Get and possibly set the system date and time from a remote host
- readahead - Pre-load files in RAM cache so that subsequent reads for those files do not block on disk I/O
- readlink - Display the value of a symlink
- realpath - Return the absolute pathnames of a given argument
- reboot - Reboot the system

- renice - Change the priority of running processes
- reset - Reset the screen
- rm - Remove (unlink) files
- rmdir - Remove the directory if it is empty
- rmmod - Unload the specified kernel modules from the kernel
- route - Edit kernel routing tables
- rpm2cpio - Output a cpio archive of the rpm file
- run-parts - Run a bunch of scripts in a directory
- runsv - Run a program in a different security context
- runsvdir - Start and monitor a service and optionally an appendant log service
- rx - Receive a file using the xmodem protocol
- script - Make a typescript of a terminal session
- sed - Stream editor for filtering and transforming text
- seq - Print numbers from first to last, in specified increments
- setarch - Change the reported architecture
- setconsole - Redirect system console output to a device(default: /dev/tty)

- setlogcons - Redirect the kernel output to a specified console
- setsid - Run program in a new session
- sh - Print or check SHA1 checksums
- sha1sum - Print or check SHA1 checksums
- slattach - Attach network interfaces to serial lines
- sleep - Delay for a specified amount of time
- softlimit - Set soft resource limits, and then run program
- sort - Sort lines of text
- split - Split a file into pieces
- start-stop-daemon - Start and stop system daemon programs
- stat - Display file (default) or file system status
- strings - Display printable strings in a binary file
- stty - Change and print terminal line settings
- sum - Checksum and count the blocks in a file
- sv - Control services monitored by runsv supervisor
- svlogd - Read log data from standard input, optionally filter log messages, and write the data to one or more automatically rotated logs
- swapoff - Stop swapping on a device
- swapon - Start swapping on a device

- sync - Write all buffered file system blocks to disk
- syslogd - System logging utility
- tac - Concatenate files and print them in reverse
- tail - Print last ten lines of each file to standard output
- tar - Create, extract, or list files, from a tar file
- taskset - Set or get CPU affinity
- tcpsvd - Create TCP socket, bind it to ip:port, and listen
- tee - Copy standard input to teach file, and also to standard output
- telnet - Connect to telnet server
- telnetd - Handle incoming telnet connections
- test - Check file types, compare values, and so on. Return a 0/1 exit code.
- tftp - Transfer a file from/to the TFTP server
- tftpd - Transfer a file on the TFTP client's request
- time - Run programs and summarize system resource usage
- top - Provide a view of process activity in real time
- touch - Update the last-modified date on the given files

- tr - Translate, squeeze, and/or delete characters
- traceroute - Trace the route to host
- true - Return an exit code of 'true'
- tty - Print filename of standard input's terminal
- ttysize - Print dimensions of standard input's terminal
- udhcpc - A very small DHCP client
- udpsvd - Create UDP socket, binds it to ip:port, and wait
- umount - Unmount file systems
- uname - Print system information
- uncompress - Uncompress .Z files
- unexpand - Convert spaces to tabs, writing to standard output
- uniq - Discard duplicate lines
- unix2dos - Convert a file from UNIX to DOS format
- unlzma - Uncompress a file
- unzip - Extract files from zip archives
- uptime - Display the time since the last boot
- usleep - Pause for a specified number of microseconds
- uudecode - Uudecode a file

- uuencode - Uuencode a file to stdout
- vi - Edit a file
- watch - Execute a program periodically
- wc - Print line, word, and byte counts for each file
- wget - Retrieve files via HTTP or FTP
- which - Locate a command
- whoami - Print the username associated with the current effective user ID
- xargs - Execute a command on every item given by standard input
- yes - Output a string repeatedly until killed
- zcat - Uncompress to stdout
- zcip - Manage a ZeroConf IPv4 link-local address

BusyBox Command Limitations

The following commands are removed from the busybox configuration and do not apply to the Intel® CE Media Processor CE 4100 Development Platform:

- adduser - Add a user
- chgrp - Change the group membership of each file to group
- chown - Change the owner and/or group of each file to owner and/or group

- chvt - Change the foreground virtual terminal to /dev/ttyn
- crond - Daemon to execute scheduled commands
- delgroup - Delete a group from the system or a user from a group
- deluser - Delete a user from the system
- dumpkmap - Print a binary keyboard translation table to standard output
- fbset - Show and modify frame buffer settings
- fdflush - Force floppy disk drive to detect disk change
- fdformat - Format floppy disk
- fsck - Check and repair file system
- fsck.minix - Check MINIX file system
- httpd - Listen for incoming HTTP requests
- ifdown - Take down a network interface
- ifup - Bring up a network interface
- iptunnel - iptunnel {add | change | del | show} [NAME]
- kdb_mode - Report or set the keyboard mode
- loadfont - Load a console font from standard input
- loadkmap - Load a binary keyboard translation table from standard input
- login - Begin a new session on the system

- logname - Print the name of the current user
- mdev - Mini-udev implementation
- mkfs.minix - Make a MINIX file system
- modprobe - Add or remove modules to or from the Linux kernel
- mt - Control magnetic tape drive operation
- passwd - Change the user's password
- pivot_root - Move the current root file system to and make a new root file system
- pscan - Scan a host and print all open ports
- raidautorun - Tell the kernel to automatically search and start RAID arrays
- readprofile - Read kernel profiling information
- resize - Resize the screen
- rpm - Manipulate RPM packages
- runlevel - Report the previous and current runlevel
- setkeycodes - Set entries into the kernel's scancode-to-keycode map
- setuidgid - Set user ID and group ID to account's user ID and group ID
- su - Change user ID or become root
- sulogin - Single user login
- switch_root - Switch to another file system as the root of the mount tree

- sysctl - Configure kernel parameters at runtime
- udhcpd - A very small DHCP server
- vconfig - Create and remove virtual Ethernet devices
- vlock - Lock a virtual terminal
- watchdog - Periodically write to watchdog a device
- who - Show who is logged on

III. Hooking up to the Development Platform

Setup Tools for the Development Platform

You will need several tools before trying to communicate with the platform. First, a communication application is necessary for a serial connection. When using a Linux host computer, try: **Minicom**.

If minicom is not installed, type the following in the Linux Terminal:

```
apt-get install minicom
```

When using Windows, however, either of the following will work:

- TerraTerm
- HyperTerminal

Second, you will need an application that can automatically authorize any FTP requests. This is will make easier to send data/files to and from the Platform.

- Pure Admin

Serial Port to USB/Serial Port Information

Remoting into the Videon Central AV platform box is simple. A female RS-232 cable is needed to connect the host computer and the platform. A male RS-232 to USB adapter may also be used if the host computer doesn't have a serial port. After connecting the Videon Central AV platform to the host computer, the communication application (ie Minicom) needs to be configured before you can use it.

To configure Minicom, open the Linux Terminal and type:

```
minicom -s
```

This will bring up the Minicom configuration tab. Go down to "Serial port setup" and select it. This is where the serial port can be configured on the Host PC. Change the following specifications so that they match the following:

- bit rate: 115200
- data bit: 8
- parity: none
- stop bits: 1
- flow control: none

Lastly, the Serial Device should be changed to the correct tty port. The Serial Device will allow your Host PC to see and communicate with the Videon Central AV platform. When using the serial to serial connection to the Development Platform, configure the Serial Device to /dev/ttyS0. When using a serial to USB connection, however, change the Serial Device to /dev/USB0. The device number may have to change (eg /dev/ttyS1) depending on how many serial or USB connections are being used by the Host PC.

When finished with "Serial port setup", go down and select "Save setup as..". Choose a title to name the configuration and hit Enter. For future reference, this will open a connection to the Development Platform by just typing the following in the Terminal:

```
minicom <name>
```

The command will initialize the modem with the correct settings and configurations. For now, hit Exit on the configuration tab, and this will automatically start the modem with the applied settings. At this point, the Platform's file system should be visible in the Terminal.

IV. Connecting the Videon Central AV platform to a TV display

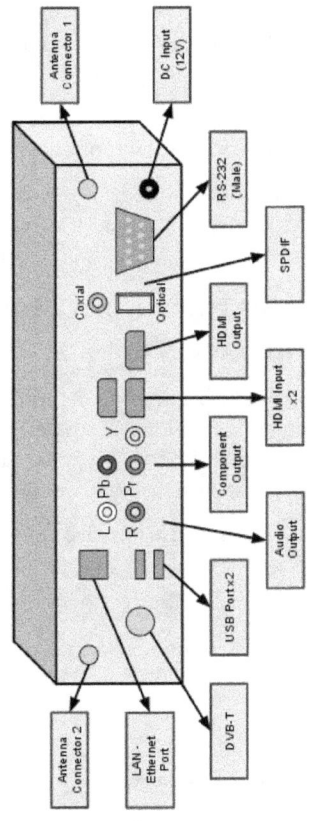

Figure 2 – Videon Central AV Platform input/output connections

Connecting the Development Platform to a TV display can be done using either of the following methods:

- **Component:** Connect the Component input on the TV to the Component outputs on the Platform.
- **HDMI:** Connect the HDMI input on the TV to the HDMI output on the Platform.
- **S-Video:** Connect the S-video input on the TV to the S-video output on the Platform.
- **Audio:** Connect the analog audio cable on the TV to the audio output jacks on the Platform.

Transferring Files from Host to Target and Vice Versa

ftpget - Allows you to retrieve a remote file on a host computer using FTP. Note: Transferring directories must be compressed first. To extract a compressed directory on the Platform, use the tar command.

Usage: `ftpget [options] remote-host local-file remote-file`

Options:

    -c,--continue   Continue previous transfer

    -v,--verbose   Verbose

    -u,--username Username

    -p,--password Password

    -P,--port   Port number

ftpput - Allows you to store a local file on the host computer using FTP. Note: Transferring directories must be compressed first. To extract a compressed directory on the Platform, use the tar command.

```
Usage: ftpput [options] remote-host
remote-file local-file
```

Options:

    -v,--verbose   Verbose

    -u,--username Username

    -p,--password Password

    -P,--port   Port number

wget - Allows you to retrieve files using either HTTP or FTP

```
Usage: wget [-c|--continue] [-s|--spider]
[-q|--quiet] [-O|--output-document fi]
 [--header 'header: value'] [-Y|--proxy
on/off] [-P DIR]            [-U|--
user-agent agent] url
```

Options:

-s	Spider mode - only check file existence	
-c	Continue retrieval of aborted transfer	
-q	Quiet	

-P   Set directory prefix to DIR

-O   Save to filename ('-' for stdout)

-U   Adjust 'User-Agent' field

-Y   Use proxy ('on' or 'off')

Helpful Platform Commands

This section will briefly touch on a few simple but important commands and that will enable features on the Platform. For more useful commands, see Sample Scripts.

chmod

> This changes the permission of any file. If a file requires root privileges, chmod would come in handy here.

udhcpc

> This enables networking for FTP, and is needed for functions such as ftpget, ftpput, wget, etc.

tvmode

> This initializes GDL, which allows most applications and samples to run. The default output on the Platform is component.
>
> ```
> /bin/gdl_samples/portattrs -port 2 -set 0 0
> ```
>
> This enables HDMI output on the Platform

V. Setting Up and Understanding the SDK

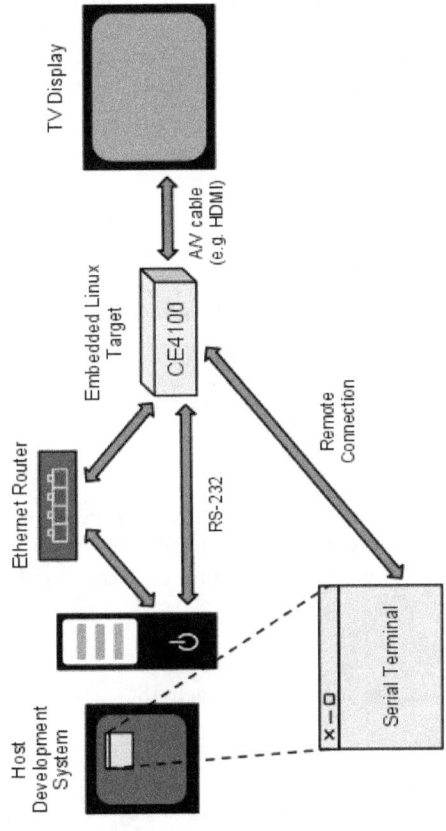

Figure 3 - Development Environment Overview

Figure 3 is how the development should be set up to continue setting up and understanding the SDK.

It is very **important** that the BASE ISO is installed first, otherwise running "make all" will encounter errors. The BASE ISO is meant to be the platform upon which other ISOs can provide layering support. For extra precaution, each ISO file comes with a checksum file so the ISOs can be validated. The ISO images can be either directly mounted onto the Linux host computer or burned onto CDs to be read by an optical drive.

Installation Steps:

1) Mount the ISO using the following command:

```
sudo   mount   -t   iso9660   -o   loop,ro
<ISO>.iso /mnt
```

Your mounted ISO will now appear in the root directory under /mnt. You may also use an application such as Furius ISO Mount to mount and unmount your ISOs.

2) Run the installer for each ISO and set the install patch:

```
bash /mnt/installer /install_patch
```

When you run the install script, two SDK license agreements will appear. Accept both agreements to continue the SDK install. The script will install to a directory called "IntelCE-<version>" inside the /install_path that you specified. Do this for every ISO to complete the SDK build.

Once you have all the ISOs installed, the "IntelCE-<version>" directory should look like this:

- binaries
- build_i686
- dl
- docs
- INTELCE_LICENSES
- package
- project
- project_build_i686
- scripts
- target
- toolchain
- toolchain_build_i686
- CHANGES
- Config.in
- COPYING
- installation.log

- make
- Makefile
- skus_installed
- targetFS.log
- TODO
- version

Now open the Linux Terminal and go to your "IntelCE-<version>" directory:

```
cd /install_path/IntelCE-<version>
```

Then run the following command to configure the SDK:

```
make menuconfig
```

Select "Package selection for target" in the configuration window. Go through the list and deselect ramdisk and targetFS because they require root permission. Then exit out of the configuration to get back to the terminal.

Build the SDK by running the following command:

```
make all
```

Build ramdisk and targetFS by running the following command:

```
sudo make targetFS
```

The SDK Development Environment

Installing the SDK will setup the development environment for you to develop and compile applications and drivers on the Intel® CE Media Processor. The SDK includes a variety of libraries, source code and toolchain utilities. One of the more important directories in the SDK, is the "i686-linux-elf" directory:

Inside the SDK directory, /IntelCE-<version>, you will find another directory called /build_i686/i686-linux-elf. It includes binaries, documents, scripts, libraries, and source code for the SDK. The following will explain and detail each directory in i686-linux-elf:

/i686-linux-elf/bin

The bin directory contains applications and utilities that can configure the Development Platform, and the source code for these binary files are included in the SDK as well. All source code files can be found in the /i686-linux-elf/src directory. It is important to note that the gcc compiler and the toolchain binaries are copied into the bin directory when the SDK is installed. Both of these are needed by the makefiles.

/i686-linux-elf/lib

The lib directory includes open-source, precompiled libraries such as GStreamer and Intel® created libraries. These libraries are used by the Intel® CE Media Processor SDK. Applications or other libraries will be able to find this directory since /i686-linux-elf/lib is a known, fixed path. If you are making a new library to run on the Platform, then you should modify your makefiles to export those libraries to /i686-linux-elf/lib.

/i686-linux-elf/include

The include directory stores a copy of the header files needed for SDK.

/i686-linux-elf/src

The src directory includes tarballs for the source code for all of the SDK and open source components. These are meant to be used as a backup, while you can modify the source code in the SDK. Each tarball will have a makefile or build script that will tell you how to build the source code.

/i686-linux-elf/usr/include

The /usr/include directory is similar to the include directory as it holds C library and kernel header files.

The kernel header files are used by the Platform's applications, which are compiled by the toolchain in the SDK rather than the host computer.

/i686-linux-elf/kernel/linux-2.\<version\>/include

The `/kernel/linux-2.<version>/include` directory contains a copy of the target kernel header files.

/i686-linux-elf/libiconv-1.11.1

The `/libiconv-1.11.1` directory contains header files, binaries, and libraries for libiconv-1.11.1.

The Target File system

The following steps below will create a bootable ramdisk image that configures and runs the drivers for the CE4100 Platform. After the "make targetFS" command is done, the files in the ramdisk image can be found at `/IntelCE-<version>/project_build_i686/IntelCE/root` on the SDK. You may now send this over to the CE4100 Platform by either NFS Share, copying onto a SATA drive, or compressing the files into a ramdisk image. Note, there may be some files in the SDK that are not included by default. Additional binaries and libraries can be found here:

```
IntelCE-<version>/build_i686/i686-
linux-elf/bin
```

```
IntelCE-<version>/build_i686/i686-
linux-elf/lib
```

The ramdisk component can also create a nearly empty file system image, which you can build on. The ramdisk will have the following files created:

- ramdisk.ext2
- initrd.gz
- ramdisk.cramfs

These compressed file system images only has BusyBox utilities and boot files that will go to the Linux command-line. They do not have drivers or libraries that would enable graphics and media playback. You can either ramdisk.ext2, initrd.gz, or ramdisk.cramfs or customize your own compressed file system.

Debugging

GDB and SVEN are two different debuggers that are provided on the SDK. GDB focuses on application level support and SVEN focuses on driver level support.

SVEN (System Visible Event Nexus)

SVEN stores real-time software "event traces" and then lists these software events with accurate timestamps. Sven is the standard debug tool for the CE4100 Platform as it is built onto all media and display drivers. The scripts directory on the CE4100 Platform has example scripts of SVEN. On the target file system the /scripts/ directory contains example scripts. One of these example scripts include sven_AV_heatlh_capture.sh, which has a --sysinfo option to help log system information for bug reports.

GDB (GNU Project debugger)

GDB is the GNU Project debugger. This will allow you to see what happened to a program when it reaches an error. Remote debugging is possible through the GDB which runs on the host and the GDBserver which runs on the target platform. Both the GDB and GDBserver are provided on the SDK. The GDB is a debugger that allows you to view the status of a program or process. The GDBserver is a program that allows the GDB to run on the host while debugging a program or process on the target platform.

For more information about the GDB and the GDBsever:

http://sourceware.org/gdb/

VI. Samples, Applications, and Scripts for CE4100 Platform

Information on GDL, Graphics, DirectFB Samples

GDL Information and Samples

The SDK includes sample applications that illustrate the use of the display driver or GDL on the Intel® Media Processor. The executable sample applications are located at /bin/gdl_samples on the target file system after the SDK has been installed on the Platform. There is also a readme file in that directory, which provides a brief explanation of each sample application. The source code for each sample can be found on the SDK in the IntelCE-<version>/project_build_i686/IntelCE/display-<version>/src/apps/samples/src/ directory.

The display driver consists of several components including a kernel module called gdl_mm.ko for inter-process communication and interrupts, a Daemon called gdl_server for controlling the video display, and a number of shared objects.

Load the kernel module:

```
/sbin/insmod /lib/modules/gdl_mm.ko
```

Start the daemon server:

```
/bin/gdl_server
```

A brief description of each GDL sample:

Sample	Description
cfgplane	Allows user to configure any attributes of a plane
color_bars	Color bars are displayed on screen for testing
hdmi_sink_info	Queries the capabilities of an HDMI-connected sink
list_modes	Outputs a list of supported display modes
plane_attr	Queries the current settings of a plane's attributes
pngview	Displays a .png file on a selected plane
portattrs	Reads and sets port driver attributes

Graphics Information and Samples

The SDK provides several GFX samples, which illustrate the use of the graphic capabilities of the Platform such as OpenGL ES, OpenVG, EGL and SRB. The executable sample applications are located at /bin/gfx_samples on the target file system after the SDK has been installed. The graphics driver configuration file can be found in /etc/powervr.ini. A sample-description file and the source code can be found on the SDK in the IntelCE<version>/project_build_i686/IntelCE/graphics-<version>/src/gfx_samples.tgz directory.

The graphics driver includes a kernel module called pvrsrvkm.ko for communication with graphics core as well as a number of shared objects. The graphics driver initialization script can be found in /etc/init.d/graphics. The kernel module, pvrsrvkm.ko, must initialized for the graphics engine to work:

/sbin/insmod /lib/modules/pvrsrvkm.ko

Both the start up script and the kernel module should be initialized upon boot up.

Below is a brief description of each Graphic sample:

Sample	Description

Bounce	Renders a bouncing square
Cube	Renders a rotating cube
bounce2	Renders a bouncing square
cube2	Renders a rotating cube
sphere2	Renders a sphere
Bouncevg	Renders a bouncing rectangle inside another rectangle
water	Shows a two-dimensional water wave simulation
showcfgs	Lists attributes of one or all EGL configurations
matching_config	Shows how to select an EGL configuration
cached_pixmap	Shows how to use eglWaitNative to sync with CPU access
bind_pixmap	Uses EGL_INTEL_bind_pixmap extension
bind_pixmap2	Uses EGL_INTEL_bind_pixmap

	extension
egl_image	Uses EGL_KHR_image extension
egl_image2	Uses EGL_KHR_image extension
egl_image_memory	Uses EGL_KHR_image and EGL_INTEL_image_memory extensions
egl_image_memory2	Uses EGL_KHR_image and EGL_INTEL_image_memory extensions
bounce_srb	SRB sample. Blits and fills a bouncing square
vidtex	Renders a video stream as a texture on rotating geometry.
vidtex2	Renders a video stream as a texture on rotating geometry.

DirectFB Information and Samples

The SDK includes several samples applications that illustrate the use of DirectFB on Intel Media Processors. The CE4100 Platform comes with DirectFB version 1.4.3 and the configuration file can be found in `/etc/directfbrc`. The executable sample applications are located in `/usr/local/bin` on the target file system after the SDK has been installed. The source code can be found on the SDK in the `IntelCE<version>/project_build_i686/IntelCE/DirectFB_<version>/src/apps/samples` directory.

Before these samples can be run, GDL must first be initialized. This can be done by calling the tvmode command (see Sample Scripts). DirectFB also depends upon the Fusion kernel and number of shared objects, which should be initialized on boot up. The following command should be able to run the Fusion kernel module:

```
insmod /lib/modules/fusion.ko
```

Below is a brief description of DirectFB samples (distributed by DirectFB):

- df_andi
- df_dok
- df_drivertest
- df_fire

- df_fonts
- df_input
- df_knuckles
- df_layer
- df_neo
- df_porter
- df_texture
- df_window
- spacedream

Below is a brief description of DirectFB samples (distributed by Intel®):

Sample	Description
alpha_layer	A pre-computed image that is blitted onto the alpha layer
alpha_premult	Shows two display layers that should match when compared. Once uses pre-multiplied alpha color components and the other uses non-pre-multiplied alpha color components.
bg_color	Changes the background color of Mixer A

bouncing_images	Two display layers are reduced in size then images blit onto them.
deferred_rendering	Demonstrates the difference between GPU-based deferred rendering and immediate rendering of a graphics drawing operation.
encoder	Shows the output display resolution using encoders.
encoder_attributes	Changes the attributes of the component encoder.
encoder_signal	Changes the component encoder from one mixer to another then turns the output signals of all three encoders on and off.
idirectfbgl_example	Demonstrates using IDirectFBGL and DirectFBIntelCE interfaces.
idirectfbgl_example2	Demonstrates using IDirectFBGL and DirectFBIntelCE interfaces.
layer_opacity	Draws two rectangles and changes their opacity with each key press.

layer_src_color_key	Changes the hardware source color key of the display driver
layer_visibility	Demonstrates how to turn layers on and off using mixers
multiAppTestTool	Sample tool used for demonstrating and testing the multiple DirectFB application usage model.
TestLayers	Tests all types of DirectFB display layers and their capabilities.

Information on Media Applications

There are several SMD applications that are meant to demonstrate a variety of media playback and audio features. The source code for these applications are located in the

`IntelCE<version>/project_build_i686/IntelCE/smd_sample_apps-linux-x86` directory on the SDK.

Common Utility Functions:

cl_app - cl_app streams audio/video files and system-level features. This program relies on configuration files, which allows the application to support various media types - see Sample Configuration Files. This application uses the display driver, ISMD modules, and system infrastructure to playback media files or through a Transport Stream.

The cl_app application includes the following features:

- Media file playback
- Tuner playback
- Configurable pipeline modes
- PSI parsing for transport streams
- MPEG-2 transport streams, Blu-Ray/DLNA transport streams (no AACS), DVD program
- frame step, trick modes, streams, pause/resume
- pip and multistream
- channel-change
- DVR included

The following is a list of command-line and run-time commands for cl_app:

- cl_app <input_filename>: starts auto playback of specified file
- cl_app @TSIN=X: starts auto playback from TSIN number X

- cl_app <config_filename>: starts playback, using specified config
- cl_app <input_filename> <config_filename> : starts auto playback of specified file or TSIN, using specified config
- cl_app <config(0)> <config(1)> ... <config(n)>: starts multi-stream playback, using specified configs
- cl_app <input_filename> <input_filename> ... <input_filename>: starts multi-stream playback, using specified inputs and default multi-stream configuration

run-time commands:

- 'x, q' - exit (quit) the application.
- 'p' - pause playback.
- 'r' - resume playback (unpause, or exit trick modes).
- 'spacebar' - pause toggle.
- 'l' - toggle playlist looping.
- 'up arrow' - skip to next program/stream.
- 'down arrow' - skip to previous program/stream.
- 'h' - Display help information.
- 'c' - Display current configuration.
- 'd' - Toggle deinterlacing enable.
- 'v' - Display current video content properties.

- 'a' - Display current audio content properties.
- 'i' - Display current system info statistics.
- 'm' - Mute (disable) audio input.
- 'b' - Mute (disable) video input.
- 'mp' - Mute entire audio processor.
- 's' - Loop through scaling policies.
- 'g' - Toggle gaussian enable.
- 'k' - in multistream mode, pip-swap (focus change)
- 'dr' - Toggle de-ringing filter enable.
- 'dr' <level> - Set de-ringing filter to level [1-100].
- 'tm' - Interactive trick modes.
- 'atm' - Advanced interactive trick modes.
- 'oob' - Out-Of-Band (dynamic) trick modes.
- 'fix' - Set/disable fixed-framerate
- 'pos' - Current stream position.
- '<number>' - Select stream to direct commands to.
- '<number>' - Select stream to direct commands to.
- 'r arrow' - (when unpaused) switches to next trick mode.
- 'l arrow' - (when unpaused) switches to prev trick mode.
- 'tag' - Insert a client ID tag (pull mode only).

- ``audout' - Interactive dynamic audio output configuration.
- ``adv' - frame step, can be used in forward or reverse playback.

sg_player - sg_player demonstrates that GStreamer elements can be interfaced with the SMD drivers. This program relies on configuration files, which allows the application to support various media types. See Sample Configuration Files.

The sg_player application includes the following features:

- It demonstrates how to use ISMD GStreamer elements to playback a media file.
- It demonstrates the beginning of 3rd party soft GStreamer element integration with ISMD GStreamer elements to extend its functionality (e.g. soft MPEG2 video decoder, soft AC3 audio decoder, and soft PS Demux element). See the ISMD release notes for more details on the ISMD GStreamer Plugin.
- It supports audio video playback of MPEG-2 Transport streams and Program streams
- It supports the video formats of MPEG2, H.264, and VC-1.

- It supports video output to component and HDMI.
- It supports the audio formats of PCM/LPCM, AC3(DD), MPEG1-2 (MP2), MPEG1-3 (MP3), MPEG2/4 AAC (ADTS only), WMA (WMAv2 stereo-only), and DTS
- It supports audio output to I2S0, I2S1, SPDIF, and HDMI ports
- The sg_player is written in C.
- For a detailed description of ISMD GStreamer elements used in sg_player, please refer to the Intel Media Processor CE 3100 GStreamer SMD Programming Manual.
- For details of GDL and other non-SMD features/APIs, please refer to GDL and other related programming guides in this SDK.

The following is a list of command-line and run-time commands for sg_player:

- 'sg_player <config_filename>' - starts playback, using specified config. The configuration file must have an extension of '.cfg'.

run-time commands:

- '2','4', or '8' - basic fast-forward trick modes
- '1' - normal 1x playback

- '0' - seek back to beginning of stream (no pipeline teardown)
- 'k' - seek forward 2 seconds or 2MB in stream
- 'l' - seek forward 10 seconds or 20MB in stream
- 'j' - seek backward 2 seconds or 2MB in stream
- 'h' - seek backward 10 seconds or 20MB in stream
- '\' - query source byte position
- '->' arrow - switches to next advanced trick mode.
- '<-' arrow - switches to prev advanced trick mode.
- 'T' or 't' - change trick_mode type
- 'P' or 'p' - pause
- 'R' or 'r' - resume from pause or reset from stop
- 'Q' or 'q' - query current stream position from renderers
- 'S' or 's' - stop (pipeline unlinks)
- 'X' or 'x' - exit

audio_player - audio_player demonstrates a more broader view of the audio only playback features on the Platform; more so than cl_app. This program relies on configuration files, which allows the application to support various media types. See Sample Configuration Files. Audio_player uses ISMD modules, and system infrastructure to playback media files or through a Transport Stream.

The audio_player application includes the following features:

- file playback
- ability to configure physical audio inputs/outputs
- ability to add multiple audio inputs/outputs
- capture processed audio to an output file
- audio per-input mixing configuration via a config file
- pause/resume
- MPEG-2 transport streams, 188-byte and 192-byte transport streams (for audio
- playback only)
- Program stream audio-only playback

The audio_player can have up to two arguments; the configuration file and the run-time commands. Below is a list of the run-time commands:

- 'p' - Pause all inputs
- 'r' - Resume all inputs
- 'f' - Flush all inputs
- 'm' - Toggle Mute

- 'n' - Go to next input in playlist
- 'e' - Reconfigure input 0 to a new file source
- 's' - Print all input stream information
- 'a' - Print stream info calling each function individually
- 'i' - Enable/Disable inputs
- 'c' - Get current stream position
- 'b' - Enable/Disable bass management
- 'right arrow' - Fast Forward
- 'left arrow' - Reverse
- '+' -Increase volume on input 0
- '-' - Decrease volume on input 0
- 'x', 'q' - Exit application

tsout_streamer - tsout_streamer utilizes a Timed Transport Stream (TTS or M2TS) over the TSOut interface using the TSOut SMD driver. This sample application shows a series of API calls that an application needs to perform in order to utilize the TSOut feature. This program relies on configuration files, which allows the application to support various media types.

The following is a list of command-line commands for tsout_streamer:

- tsout_streamer [-h] [-v] -i <file> -t <192 | 188> [-p <PCR PID>]
- -h   Help. Prints this usage message.

- -v   Verbose mode. Prints more information at runtime.
- -i   Specifies that 'file' should be used as the input source.
- -t   Specifies the Transport Stream Type.
  - 192 is the m2ts/TTS stream with 192 byte TS packets.
  - 188 is the regular transport stream with 188 bytes TS packets.
- -p   Specifies the PID in the stream that carries the PCR. Required for TS 1.

- Use the Keys x or q to exit the application
  - The Program uses the following error codes:
- 0   Success / No Errors
- -1   Generic Failure. Error message will be displayed on the console
- -2   Bad Parameters / insufficient number of Parameters provided

Sample Configuration Files for Media Applications

Each sample application requires a configuration file in order to run a media file. Configuration files are text files that specify certain options on what or how to play. These configuration files must have an extension of '.cfg'. There is an example configuration file in the bin directory called 'full_config_fle.cfg', so you can understand the default options and structure. Below is a list of configuration files that are used by their respected applications:

cl_app:

Configuration File	Description
2x_TM_Smooth.cfg	This configuration shows some of the basic trick mode options available. It is currently set to play at 2x, with all frames being displayed.
3d_mvc_config.cfg, 3d_mvc_ssif.cfg	This is a standard configuration, except that it sets the pipeline in 3D mode.
dvd_playlist.cfg	This simply shows how to construct a simple playlist. More options can be added to each

	input if desired.
dvd_program_stream.cfg	For manually playing program streams, these are the relevant options. This "manual mode" is useful if more control over the streams played is needed. This config works with AC3 program streams.
flush_channel_change_ts.cfg	To perform stream changes without tearing down and rebuilding the pipeline, the needs_new_pipe option and the needs_flush option are changed. This configuration will channel-change to the same input, more inputs can be added to introduce some variety.

pip0.cfg	This is part of a two-configuration example for dual-stream. This is the main stream. This is mostly a standard configuration, except that it sets the display plane to UPP_B.
pip1.cfg	This is part of a two-configuration example for dual-stream. This is the secondary stream. This configuration sets the display plane to be smaller, disables the display programming (it is done as part of the primary stream), and enables the viddec downscaler to decrease the performance burden on the

	vidpproc.
play_and_save.cfg	With this configuration, the data being played can also be saved to a SPTS file. It is most practical to use when playing a TSIn stream, but it will work with file playback too.
play_from_50MB_to_100MB.cfg	cl_app can play portions of a file, using the file_start_offset and file_end_offset options. These are in bytes. Note that this must be used with a file that is at least as large as the end offset.
transport_stream_manual.cfg	Generally, transport streams do not need to have media_info specified as the PI

	parsing will gather that information. This simply shows how one can disable PSI parsing and manually specify the PID values and algorithms.
ts_in_save_no_play.cfg	To save data from tuner inputs (TSI), all that is needed apart from normal playback is to specify the INPUT_FILE pipeline mode, which saves the SPTS input from the TSIN directly to a filesink.
tsi_pip_primary_mspod.cfg, tsi_pip_secondary_mspod.cfg	To use the tuner inputs (TSI) in MSPOD mode, all that is needed in addition to normal TSI playback is the specification of the

	ms_pod_mode mode in the inputs section. This is part of a dual-stream "pip" configuration, although can be used on its own as well.
ts_in.cfg	To use the tuner inputs (TSI), all that is needed is for the input_source block to have tsi_interface specified. Note that an input filename should not be present in this configuration.

sg_player:

Configuration File	Description
sg_player_sample_config.cfg	Sample configuration file for sg_player

audio_player:

Configuration File	Description
audio_player.cfg	N/A
audio_quality.cfg	Use this config file with audio_player to build up an audio control and pipline for a given output. This file should contain one instance of "audio_quality_pipe" and may contain multiple control and filter instances
input_mixer_config.cfg	This is the format of a config file for each input

Device Memory Modification for Display/SMD

Device and kernel memory is allocated during system boot. The amount of memory reserved for devices is 768MB by default with a base address or memory shift of 256MB. The first 256MB is used for the Linux kernel. The user can alter and configure the memory allocation in the platform configuration file. This file can be found in the /etc/platform_config/memory_layout_<system memory value>.hcfg file.

Example, to change the configurations on one of the SMD properties, the line would look something like this:

```
smd_buffers_ALSA         { buf_size =
0x2000    base = 0x00520000    size =
0x000400
```

Before changing the configuration file, there are a few variables that you should know:

- **buf_size** - specifies the size of individual buffers allocated from this section of memory
- **base** - specifies the physical address of the start of this section of memory
- **size** - specifies the total number of reserved bytes

VII. Script Information and Samples

Any scripts that are made inside the Development Platform should ideally be placed inside the scripts directory. These scripts can be called anywhere inside the Platform console as long as the path is setup.

To make a new path, type the following:

```
export PATH=$PATH:/<path to script>
```

To see existing paths, type the following:

```
$PATH
```

This book includes six scripts that will help get you started using and developing your applications quickly. You can download these scripts at:

Here is a list of sample samples; feel free to use them:

1) **start_script** – *enables the following important features of the Videon box*

```
udhcpc //enables networking for FTP

tvmode  1920x1080p59.94   //starts  gdl  and
enables display output

/bin/gdl_samples/portattrs -port 2 -set 0 0
//enables hdmi output
```

2) **GetFile_script** - *gets file from host computer and transfers to development platform*

```
echo "getting  $1  and  saving  to  $1"
//displays file being transferred
```

```
ftpget -u <host name> -p <password> <IP
Address> $1 $1 //FTP command

chmod 777 $1 //changes permissions for
complete control
```

3) **PutFile_script** - *gets file from development platform and transfer to host computer*

```
echo "getting $1 and saving to $1"
//displays file being transferred

ftpput -u <host name> -p <password> <IP
Address> $1 $1 //FTP command
```

Important These next few scripts must written as follows: ". <script>". Otherwise, the script will run in it's own shell and will not perform as expected.

4) **.GDL_script** - *changes directory to GDL samples*

```
cd bin
cd gdl_samples
echo ">>> GDL samples! <<<"
ls
```

5) .DirectFB_script - *changes directory to DirectFB samples*

```
cd usr
cd local
cd bin
echo ">>> DirectFB samples! <<<"
```

ls

6) **.GFX_script** - changes directory to GFX samples
cd bin
cd gfx_samples
echo >>>GFX Samples!<<<
ls

VIII. SDK Development Setup and Custom Applications

SDK Development Setup

One of the more important aspects of the Development Platform is the ability to build and compile your own applications and run them on a TV display through the CE4100. This section will be divided in two parts: applications that run in the CE4100 console/terminal and applications that run on a TV display. Before anything else, you need to setup your SDK Development Environment. Note that all script and make files can be downloaded at http://www.glassthumb.com/wiki/ce4100pocketguide.

In order to setup your SDK environment, go to your host computer and run the following commands directly in the terminal:

Note: If you run this as a script, the commands will run within its own shell, and will not work as intended.

```
export CANMORE_SDK=(<root
directory>/IntelCE-
<version>/build_i686/i686-linux-elf)
export BUILD_DEST=(<user directory>)
mkdir -p ${BUILD_DEST}/lib/pkgconfig
```

```
export CC=$CANMORE_SDK/bin/i686-cm-
linux-gcc
export CXX=$CANMORE_SDK/bin/i686-cm-
linux-g++
export AR=$CANMORE_SDK/bin/i686-cm-
linux-ar
export LD=$CANMORE_SDK/bin/i686-cm-
linux-ld
export M4=$CANMORE_SDK/bin/i686-cm-
linux-m4
export NM=$CANMORE_SDK/bin/i686-cm-
linux-nm
export RANLIB=$CANMORE_SDK/bin/i686-cm-
linux-ranlib
export BISON=$CANMORE_SDK/bin/i686-cm-
linux-bison
export STRIP=$CANMORE_SDK/bin/i686-cm-
linux-strip
export LDFLAGS="-L${BUILD_DEST}/lib -
L${CANMORE_SDK}/usr/lib -
L${CANMORE_SDK}/usr/local/lib -
L${CANMORE_SDK}/lib"
export /
LD_LIBRARY_PATH="${BUILD_DEST}/lib:${CA
NMORE_SDK}/usr/lib:${CANMORE_SDK}/usr/l
ocal/lib:"
export
PATH="${BUILD_DEST}/bin:/bin:/usr/local
/bin:/usr/bin:/sbin"
export
PKG_CONFIG_PATH="${BUILD_DEST}/lib/pkgc
```

```
onfig/:${CANMORE_SDK}/usr/local/lib/pkg
config/"
export FONTCONFIG_LIBS="-
L${BUILD_DEST}/lib -lfontconfig -
lfreetype -lz -lxml2 -lm"
export COMMON_CONFIG="--
prefix=${BUILD_DEST} --disable-static"
```

These commands will create the variables needed for the SDK. For the first two lines, be careful where you set the "CANMORE_SDK" and "BUILD_DEST" path. "CANMORE_SDK" should be routed to the i686-linux-elf folder in your SDK. "BUILD_DEST" can be placed anywhere, but it is recommended that you place it in your user directory. This path will be used to store custom applications. Make sure you route both paths from the root directory.

It is also important to note that you should only use the builders and compilers from the SDK and not from your Host PC. If everything above was done correctly, you do not have to worry about this.

Basic Applications Running on CE4100 Console

For applications that run on the CE4100 console, you may create any simple application in the C language. A "hello world" application was made as an example:

```c
#include "stdio.h"

void main() {
    printf("hello, world\n");
}
```

Once your source file has been made, type the following command to compile your application:

```
${CC} ${LDFLAGS} -o
<$BUILD_DEST/exectuable name> <source file>.c
```

This will create an executable file that can be run from the terminal. The executables will be stored in your "BUILD_DEST" directory. This executable can then be transferred over to the CE4100 Platform using, for example, ftpget (see Script Information and Samples). Now you will be able to run your simple application on the Development Platform.

Advanced Applications Running on TV Display

For applications that run on a TV display, makefiles from DirectFB is recommended. Applications in DirectFB can be more complex and advanced than console applications. You will have more options and capabilities using this method such as shapes, text, object movement, etc. For more information on DirectFB API, go to this site:

http://directfb.org/docs/DirectFB_Reference_1_5/index.html

You will need to use DirectFB makefiles in order to build and compile your applications. The DirectFB makefiles can be found in the `/IntelCE-<version>/project_build_i686/IntelCE/DirectFB_<version>/src/DiretFB-examples-<version>/src` directory. It is recommended that you use the makefiles from http://www.glassthumb.com/wiki/ce4100pocketguide since it may be easier to customize. Once you have downloaded the makefiles online, replace "Makefile" and Makefile.in" in the src directory from the files online. Before you configure these makefiles, you should create your application first.

If you are not sure what kind of application to create, there is a sample application from http://www.glassthumb.com/wiki/ce4100pocketguide that can be used for testing. This sample can be compiled using the DirectFB makefiles and can use DirectFB on the CE4100 Platform to run. This sample application will show a green "hello world" moving across the screen. It will demonstrate some of the features from DirectFB API such as:

- Font Text
- Font Color
- Object movement

Note: If you want to use text in your application, then you will need a .ttf file. A sample can be found at http://www.glassthumb.com/wiki/ce4100pocketguide. The source code to use the .ttf file can be seen in the sample application above.

You should store your application in the src directory. Once this is done, you are now ready to configure the makefiles. You don't have to worry about the variable paths in the makefiles as that has been taken care of during the SDK installation. However, you will have to add a few lines of codes to the makefiles in order to compile your application. Follow the steps below:

- Open "Makefile" in the src directory
- There will be four "checkpoint(s)" where you need to add your apps name. Search for "checkpoint" in the document and add your applications name where specified. More instructions may be found in the checkpoint comments. You may add multiple applications to this document without a problem.
- Now save the document.
- Do the same for "Makefile.in"

Once you have made your application in the src directory and configured the makefiles, the application can then be compiled. Go to the Linux Terminal, and navigate to the DirectFB src directory or to where your application source file is located. Then run the following command:

```
make
```

If you application had no errors, then you should have an executable for your application. Transfer over your application over to the CE4100 Platform using, for example, ftpget (see Script Information and Samples). Now you will be able to run your simple application on the Development Platform.

IX. RedBoot and CEFDK Shell

Redboot Introduction

Redboot stands for Red Hat Embedded Debug and Bootstrap firmware. It is used on a variety of different embedded operating systems, and provides a standard debug and bootstrap environment for the user. In order to get into the Redboot command console, press control+C when prompted as the platform begins to initialize. The RedBoot shell provides significant information that can be utilized in a number of ways. These utilities include status and troubleshooting, managing configuration information in flash, and, depending on how the boot script is setup, can boot the SDK kernel, root file system, and/or Ramdisk images from SATA, TFTP/NFS, and Flash.

CEFDK Shell and Commands

Before looking at the Reboot commands and scripts, you should be aware of the CEFDK shell. The CEFDK shell provides its own list of system boot and memory related commands. You can enter the CEFDK shell by hitting 'enter' during system boot up before Redboot initializes.

Here is a list of CEFDK commands:

- bootata - Boots from the primary master ATA device.
- bootlinux - Boots Linux from flash. Usage: bootlinux "<kernel cmd line>"
- ymodem - Receive a file from serial using YMODEM.
- lspci - Displays PCI device info.
- ord[2|4] - Read or write to memory.
- pci[2|4] - Read or write to PCI configuration space.
- port[2|4] - Read or write to I/O port.
- mmap - Displays a system memory map.
- expflash - Access flash on expansion bus.
- nandFTLL - Access NAND flash via FTL-Lite API.
- bootflash - Boot redboot from NOR or NAND flash.
- bootkernel - Boot Linux kernel from NAND flash.
- nandTest - Test nand flash compatibility
- md5 - Calculate a MD5 sum for an input data string.
- sha - Calculate a SHA sum for an input data string.
- setFPA - Enable or disable fast path audio.
- ata-map - Sets the ATA geometry mapping.
- help - Displays this screen.
- exit - Stops the shell.

Redboot Commands

Here is a list of Redboot commands:

- Manage aliases kept in FLASH memory
  - alias name [value]
- Set/Query the system console baud rate
  - baudrate [-b <rate>]
- Display/switch console channel
  - channel [-1 | <channel number>]
- Compute a 32bit checksum [POSIX algorithm] for a range of memory
  - cksum -b <location> -l <length>
- Display disks/partitions.
  - disks
- Display (hex dump) a range of memory
  - dump -b <location> [-l <length>] [-s] [-1 | -2 | -4]
- Execute a Linux image
  - exec [-w timeout] [-b <base address> [-l <image length>]]
    - [-r <ramdisk addr> [-s <ramdisk length>]]
    - [-c "kernel command line"]
- Manage FLASH images
  - fis {cmds}

- Manage configuration kept in FLASH memory
    - fconfig [-i] [-l] [-n] [-f] [-d] | [-d] nickname [value]
- Execute code at a location
    - go [-w <timeout>] [-c] [-n] [entry]
- Help about help?
    - help [<topic>]
- Display command history
    - history
- Set/change IP addresses
    - ip_address [-b] [-l <local_ip_address>[/<mask_len>]] [-h <server_address>]
- Load a file
    - load [-r] [-v] [-h <host>] [-p <TCP port>][-m <varies>]
        - [-b <base_address>] <file_name>
- Compare two blocks of memory
    - mcmp -s <location> -d <location> -l <length> [-1|-2|-4]
- Copy memory from one address to another
    - mcopy -s <location> -d <location> -l <length> [-1|-2|-4]
- Fill a block of memory with a pattern
    - mfill -b <location> -l <length> -p <pattern> [-1|-2|-4]

- Network connectivity test
    - ping [-v] [-n <count>] [-l <length>] [-t <timeout>] [-r <rate>]
        - [-i <IP_addr>] -h <IP_addr>
- Reset the system
    - reset
- Display RedBoot version information
    - version
- Display (hex dump) a range of memory
    - x -b <location> [-l <length>] [-s] [-1|-2|-4]

Modifying the Redboot Script

In order to change the boot script, type this in the Redboot shell:

```
fconfig
```

This will allow you to change some of the configurations in Redboot, and one of them will be the boot script.

Script for Booting Root File system and Kernel from SATA:

Copy the following command into the boot script:

```
>> load -v -r -m disk -b 0x200000 hda2:bzImage
>> exec -b 0x200000 -l 0x300000 -c "console=ttyS0,115200 root=/dev/sda3 rw mem=exactmap memmap=640K@0 memmap=255M@1M"
```

Depending on where your file system is located, root=/dev/sda3 might need to be changed to, for example, root=/dev/sda2.

Script for Booting Kernel from TFTP and Root File system from NFS:

In order to boot from TFTP/NFS, you will need to place the kernel into a tftp server and also put the root file system onto a NFS share. You will need an Ethernet cable attached to your Platform from your network. The platform needs to be either configured to have a static IP or it needs to have access to the NFS Host through a DHCP. Once this has been done, enter the CEFDK shell and then type:

```
bootflash NOR 0xC010000
```

Now you need to configure the boot script in Redboot so that it can find the TFTP server and NFS share:

```
>> load -v -r -m tftp -h <tftp_server_ip> -b 0x200000 bzImage
>>exec -b 0x200000 -l 0x300000 -c "console=ttyS0,115200 root=/dev/nfs
nfsroot=<nfs_server_ip>:<nfs_share_path>,nolock rw ip=dhcp mem=exactmap
memmap=640K@0 memmap=255M@1M"
```

Script for Booting Kernel and Ramdisk from Flash:

Users can also write bzImage and ramdisk to Flash in RedBoot and boot the kernel and file

system directly from flash.

Depending on the flash setup, the commands will vary for both NAND and NOR Flash. First, the bzImage and ramdisk need to be burned on to Flash. The following commands are executed in the Redboot shell.

Burn bzImage onto NOR Flash:

```
>>fis init
>>load -v -r -m tftp -h <tftp_server_ip> -b 0x200000 bzImage
>>fis unlock -f 0xC0180000 -l 0x200000
>>fis create bzImage -f 0xC0180000 -b 0x200000 -l 0x200000 -e 0x200000
>>fis list //This command is to check flash map.
```

Burn bzImage onto NAND flash:

```
>>fis init
>>load -v -r -m tftp -h <tftp_server_ip> -b 0x200000 bzImage
>>fis unlock -f 0x0180000 -l 0x200000
>>fis create bzImage -f 0x0180000 -b 0x200000 -l 0x200000 -e 0x200000
>>fis list //This command is to check flash map.
```

Burn ramdisk onto NOR Flash:

```
>>load -v -r -m tftp -h <tftp_server_ip> -b 0x800000 initrd.gz
>>fis unlock -f 0xC0500000 -l 0x300000
>>fis create initrd.gz -f 0xC0500000 -b 0x800000 -l 0x300000
```

Burn ramdisk onto NAND flash:

```
>>load -v -r -m tftp -h <tftp_server_ip> -b 0x800000 initrd.gz
>>fis unlock -f 0x0500000 -l 0x300000
>>fis create initrd.gz -f 0x0500000 -b 0x800000 -l 0x300000
```

Now that the images have been burned, the Kernel can now Boot using this command:

```
>>fis load bzImage
>>fis load initrd.gz
>>exec -b 0x200000 -r 0x800000 -l 0x300000 -s 0x2000000 -c
"console=ttyS0,115200 root=/dev/ram0 mem=exactmap memmap=640k@0 memmap=255M@1M "
```

X. Appendix A: Host Environment Setup

Setup host environment on Fedora Core 8:

Install Fedora Core 8 and select "Software Development. Make sure to disable "SELinux" in the configuration and reboot before building the SDK. Additional packages must be installed using yum by running the following command:

```
yum install gcc doxygen patch
libtermcap-devel automake make libgnome
```

Setup host environment on Fedora Core 10:

Install Fedora Core 10 and select "Software Development. Make sure to disable "SELinux" in the configuration and reboot before building the SDK. Additional packages must be installed using yum by running the following command:

```
yum install gcc doxygen patch
libtermcap-devel automake make libgnome
```

Setup host environment on Ubuntu 9.04:

Install Ubunut 9.04 on your host computer. Additional packages must be installed using apt-get by running the following command:

```
sudo apt-get install automake autoconf
rpm patch libncurses5-dev doxygen
```

Then run the following commands in the Ubuntu bash shell:

```
ls -l /bin/bash
ls -l /bin/sh
```

These commands will tell you if bash is installed and check if your system has linked sh with bash.

If you don't have bash, then install it using apt-get. Then change the default shell to bash using the following commands:

```
sudo rm -f /bin/sh
sudo ln -s /bin/bash /bin/sh
```

XI. Appendix B: Online Resources

At http://www.glassthumb.com/wiki/ce4100pocketguide, you will find a number of files that may help you in developing the CE4100 Media Processor. These include audio/video samples to makefiles for DirectFB. The list below describes some of the items that can be found online:

- Decker.ttf - A font file for DirectFB
- dbc.c - A font sample file that uses the DirectFB driver
- helloworld.c - A simple hello world application
- Makefile - A makefile for DirectFB.
- Makefile.in - A makefile for DirectFB.
- Audio/Video Samples - m2ts stream that can play with a certain player
- Sample Scripts

XII. Glossary

AAC - Advanced Audio Coding is an audio codec that is part of the MPEG specification

AP - Application processor

API - Application Programming Interface is a set of algorithms used by an application program to request and carry out low-level services performed by the operating system

AV - Audio/Visual

BIOS - Basic Input / Output System

BPP - Bits Per Pixel

BSP - Boot strap processor

CPU - Central Processing Unit

CVS - Composite Video Signal

D3D - Direct 3D, Microsoft Application Programming Interface for graphics.

DAC - Digital-to-Analog Converter

Debug Port Connection - Used to provide control during debug and validation.

DP - Dual Processor

DSP - Digital Signal Processor

DST - Destination

DSTB - Digital Set Top Box

DVB - Digital Video Broadcasting is a set of standards that define digital broadcasting using existing satellite, cable, and terrestrial infrastructures.

DVB-S - Satellite television DVB standards

DVB-T - Terrestrial television DVB standards

DVD - Digital Versatile Disc

DVD-R - Recordable DVD

DVO - Digital Video Output

ELF - Executable and Linking Format

FPS - Frames per Second

FW - Firmware running on the decoder controller

GNU - A set of freely available software development tools

HDD - Hard Disk Drive

HDMI - High Definition Multimedia Interface. This interface is used between any audio/video source such as a DVD player or A/V receiver.

HDTV - High-Definition Television. HDTV HDTV generally has resolutions of 1,080-line interlaced (1080i) or 720-line progressive (720p)

HDVCAP - High Definition Video Capture

HW - Hardware

I/F - Interface

I2C* - Inter-IC Bus. A two-wire serial interface

IDL - Internal Driver Layer

Intel CEFDK - Intel Consumer Electronics Firmware Development Kit

LAN - Local Area Network

LOD - Level Of Detail, used in texture calculations.

MMU - Memory Management Unit

MP - Multiprocessor

MPEG - Motion Picture Experts Group. An organization that develops standards for digital video and digital audio compression

MSB - Most Significant Bit

OGL/OpenGL - Open GL application programming interface

PCI - Peripheral Component Interconnect bus is a bi-directional bus

PDG - Platform Design Guide

PIP - Picture In Picture display mode

Power - System power good state

PR - Product/Program Release

PVR/PDR/DVR - Personal Video Recorder or Personal Digital Recorder or Digital Video Recorder - an interactive TV-recording device that records programs, and is able to search shows based on type

Reset - Reset/restart state for elements within the system

RF - Radio Frequency can be directly received by the tuners of TVs or radio receivers

SDTV - Standard-Definition Television. A digital television system that displays in 480i or 480p resolution

SOC - System on chip

SRC - Source code

STB - Set Top Box. A device that effectively turns a television set into an interactive Internet device

TS MPEG-2 - Transport Stream. A sequence of 188-byte packets carrying the multi-program audiovisual data

UP - Uniprocessor

VDC - Video Display Controller

WAN - Wide Area Network

Index

ALSA, 13, 77

Atom Processor, 2, 10

Audio, 3, 11, 39, 66, 99, 100

audio_player, 66, 67, 75, 76

bin, 7, 8, 22, 30, 41, 44, 46, 48, 49, 51, 52, 53, 56, 69, 78, 79, 80, 82, 98

boot, 8, 13, 30, 49, 53, 56, 76, 88, 92, 93, 94

BusyBox, 13, 18, 31, 49

bzImage, 7, 93, 94, 95, 96

CE4100, 1, 2, 3, 5, 7, 10, 48, 50, 51, 56, 81, 83, 84, 85, 87, 99

CEFDK, 88, 93, 102

chmod, 19, 41, 79

cl_app, 7, 60, 61, 66, 69, 73

component, 41, 49, 58, 59, 64

configuration, 7, 18, 31, 36, 37, 45, 53, 54, 56, 60, 61, 62, 63, 65, 66, 67, 68, 69, 70, 71, 72, 74, 75, 76, 77, 88, 89, 91, 97

consumer electronic, 2, 4

Daemon, 32, 51

debug, 50, 88, 100

dev, 1, 3, 7, 15, 20, 21, 23, 27, 32, 37, 93, 94, 96, 98, 99

directFB, 7, 8

directfbrc, 56

DVD, 4, 8, 61, 101

Embedded Linux, 4

environment, 2, 21, 26, 46, 81, 88, 97

etc, 4, 7, 8, 19, 41, 53, 56, 76, 85

Ethernet, 3, 11, 34, 93

Fedora Core, 97

filesystem, 7, 20, 25, 26,

28, 29, 32, 33, 37, 49, 51, 53, 56, 88, 93

FTP, 21, 22, 31, 35, 39, 40, 41, 78, 79

ftpget, 21, 39, 41, 79, 84, 87

ftpput, 22, 40, 41, 79

GCC, 12

GDB, 13, 49, 50

GDL, 7, 41, 51, 52, 56, 65, 79

GFX, 7, 53, 80

Graphic, 54

GStreamer, 15, 47, 63, 64

hdisk1, 7

HDMI, 3, 10, 11, 39, 41, 52, 64, 101

Host, 1, 36, 37, 39, 83, 93, 97

HyperTerminal, 35

i686-linux-elf, 46, 47, 48, 49, 81, 83

Intel, 1, 2, 5, 7, 10, 18, 31, 47, 51, 56, 57, 64, 102

iso, 43

Kernel, 7, 8, 23, 93, 94, 96

lib, 8, 47, 49, 52, 53, 57, 82, 83

Linux, 1, 2, 4, 7, 8, 10, 12, 13, 14, 15, 16, 17, 33, 35, 36, 43, 45, 49, 76, 86, 89, 90

linuxrc, 8, 24

media application, 3, 7

memory, 8, 21, 55, 76, 77, 88, 89, 90, 91, 92

Minicom, 35, 36

mnt, 8, 15, 43

mount, 14, 25, 33, 43

Nandflash, 13

network, 3, 4, 5, 22, 23, 25, 26, 28, 32, 93

operating system, 4, 13, 88, 100

path, 19, 44, 45, 47, 78, 83, 89, 94

Platform, 1, 2, 3, 7, 9, 12, 18, 31, 35, 37, 39, 40, 41, 46, 47, 48, 50, 51, 53, 56, 66, 78, 81, 84, 85, 87, 93, 102

proc, 8

Processor, 1, 2, 3, 5, 7, 10, 14, 18, 31, 46, 47, 51, 64, 99, 100

Pure Admin, 35

ramdisk, 21, 45, 46, 48, 49, 90, 94, 95

Redboot, 7, 88, 90, 92, 94

Remote, 50

RS-232, 3, 11, 36

sbin, 8, 52, 53, 82

scripts, 7, 8, 27, 44, 46, 50, 78, 79, 88

SDK, 1, 14, 42, 44, 45, 46, 47, 48, 49, 50, 51, 53, 56, 60, 65, 81, 82, 83, 86, 88, 97

Serial port, 36, 37

sg_player, 7, 63, 64, 65, 75

share, 8, 93, 94

SMD, 60, 63, 64, 65, 68, 76, 77

source, 2, 14, 20, 25, 46, 47, 51, 53, 56, 59, 60, 65, 68, 75, 84, 86, 87, 101

SPDIF, 11, 64

src, 46, 47, 51, 53, 56, 59, 85, 86

SVEN, 49, 50

S-video, 39

sys, 8

System-on-Chip, 2

Target, 39, 48

targetFS, 45, 46, 48

terminal, 19, 20, 24, 25, 27, 28, 30, 32, 34, 45, 81, 84

TerraTerm, 35

tmp, 8

tsout_streamer, 68

TV, 2, 38, 39, 81, 84, 102

tvmode, 41, 56, 78

Ubuntu, 97, 98

udhcpc, 30, 41, 78

USB, 3, 11, 36, 37

usr, 7, 8, 47, 56, 79, 82, 83

var, 8

Video, 3, 10, 11, 39, 99, 100, 101, 102, 103

Videon Central, 2, 5

wget, 31, 40, 41

www.ingramcontent.com/pod-product-compliance
Lightning Source LLC
Chambersburg PA
CBHW072216170526
45158CB00002BA/625